MOLES

by Martha London

Cody Koala

An Imprint of Pop!
popbooksonline.com

abdobooks.com
Published by Pop!, a division of ABDO, PO Box 398166, Minneapolis, Minnesota 55439.

Printed in the United States of America, North Mankato, Minnesota

082020
012021

THIS BOOK CONTAINS RECYCLED MATERIALS

Cover Photo: Shutterstock Images
Interior Photos: Shutterstock Images, 1, 5 (bottom right), 10–11, 14–15, 17 (bottom right); blickwinkel/Alamy, 5 (top), 19; iStockphoto, 5 (bottom left), 7, 9, 13, 17 (top), 17 (bottom left); A & J Visage/Alamy, 20

Editors: Christine Ha and Brienna Rossiter
Series Designer: Sophie Geister-Jones

Library of Congress Control Number: 2019954985

Publisher's Cataloging-in-Publication Data
Names: London, Martha, author.
Title: Moles / by Martha London
Description: Minneapolis, Minnesota : POP!, 2021 | Series: Underground animals | Includes online resources and index.
Identifiers: ISBN 9781532167638 (lib. bdg.) | ISBN 9781532168734 (ebook)
Subjects: LCSH: Moles (Animals)--Juvenile literature. | Mammals--Behavior--Juvenile literature. | Burrowing animals--Juvenile literature. | Underground areas--Juvenile literature.
Classification: DDC 599.335--dc23

Hello! My name is

Cody Koala

Pop open this book and you'll find QR codes like this one, loaded with information, so you can learn even more!

Scan this code* and others like it while you read, or visit the website below to make this book pop.

popbooksonline.com/moles

*Scanning QR codes requires a web-enabled smart device with a QR code reader app and a camera.

Table of Contents

Chapter 1

Tons of Tunnels

Moles are **mammals**. They dig tunnels in the dirt. Some tunnels are **temporary**. Moles use tunnels to move through the ground. Other tunnels are moles' homes.

Watch a video here!

Moles live in Europe, Asia, and North America. They often live near fields. The soil near fields has few roots. As a result, moles can dig through it easily.

Some mole tunnels go 15 feet (4.6 m) underground.

Chapter 2

Strong Legs

Moles have short, dark fur. They have long noses. Moles use their noses to sense their **surroundings**. Small hairs along their body also help them sense things.

Learn more here!

A mole's front legs are strong. The legs have long, wide claws. The claws break up dirt as the mole digs.

They also help push dirt out of the mole's way.

A mole's front legs are turned outward. That helps the mole move a lot of dirt.

Chapter 3

Finding Food

Moles have small eyes. They cannot see well. But moles have good senses of touch and hearing. Moles **rely** on these senses to find food.

Learn more here!

Moles find their food while digging tunnels. Worms and insects make up most of their **diet**. But moles also eat parts of plants. Moles use lots of energy when they dig. So, they need lots of food.

Some moles eat their weight in food every day.

Chapter 4

Mole Homes

When making its home, a mole digs down deep. The mole pushes dirt out behind it. This dirt makes a pile by the tunnel's **entrance**.

Complete an activity here!

Below the entrance, tunnels lead to many rooms. Some rooms store food. Others are for sleeping. Each female mole makes a nest in one of the rooms. The mother raises its babies there.

Female moles have one **litter** each year. At first, the baby moles stay with their mother. They leave the nest when they are six weeks old. Moles live approximately three years.

Making Connections

Text-to-Self

Are there areas around you that would be a good home for a mole? Why or why not?

Text-to-Text

What books have you read about other animals that dig tunnels? What do those animals have in common with moles? How are they different?

Text-to-World

Moles dig easily in soil with few roots. Can you think of places where it would be hard for moles to dig?

Glossary

diet – the types of food an animal eats.

entrance – a way to get inside a place.

litter – a group of young animals born to an adult animal at one time.

mammal – a type of animal that has hair or fur and feeds milk to its young.

rely – to depend on something.

surroundings – the area and things that are all around something.

temporary – lasting only a short time.

Index

Online Resources

popbooksonline.com

Thanks for reading this Cody Koala book!

Scan this code* and others like it in this book, or visit the website below to make this book pop!

popbooksonline.com/moles

*Scanning QR codes requires a web-enabled smart device with a QR code reader app and a camera.